T0251367

FLORA OF TROPICAL EAST AFRICA

APONOGETONACEAE

K.A. LYE

(Agricultural University of Norway)

Rhizomatous glabrous monoecious or rarely dioecious herbs with tubers, usually growing submerged in fresh water or (after drying up) on wet soil. Leaves all basal and alternate, simple and usually with long petioles; blades oblong to linear. Inflorescence usually a simple or bifid spike (rarely the spike digitate and divided into 3–4 parts), at first enclosed in a thin caducous spathe, very rarely (in the South African species *A. ranunculiflorus*) the inflorescence is much abbreviated simulating a *Ranunculus* flower. Flowers bisexual or more rarely unisexual. Tepals 1–6 or absent, petal-like, often persisting in fruit. Stamens 1–6, rarely more; filaments free, filiform or flattened; anthers extrorse, 2-thecous, most often only 0.2–0.5 mm. long. Ovaries 3–8, free or slightly united near their base; each ovary superior and 1-locular, with 1–14 erect ovules borne along one side of the locule-wall or at the base of the locule. Fruit a 1–14-seeded follicle. Seeds discoidal to fusiform, straight or slightly curved, with a simple or double testa; endosperm absent.

A monogeneric family widely distributed in the tropical and subtropical parts of the Old World, but absent from the Americas.

APONOGETON

L.f., Suppl. Pl.: 32 (1781); Engl. in E. & P. Pf. II. 1: 218 (1889); K. Krause in E.P. IV. 13: 9 (1906); H. Bruggen in B.J.B.B. 43: 193 (1973) & Bibl. Bot. 33, 137: 1 (1985), *nom. conserv.*

Small to medium-sized perennial herbs growing in seasonally wet soil or submerged in temporary pools or similar habitats, rarely in permanent water; leaf-blades and flowers often floating on the surface of the water. Tuber globular to oval or more irregular in shape, usually grey to brown with very numerous roots from upper part. Leaves usually with a long petiole and a distinct blade with a few parallel primary veins and numerous transverse secondary veins, but in some species, e.g. *A. vallisnerioides*, the leaf-blade is linear. Inflorescence with spirally or unilaterally arranged flowers in a loose or dense simple or bifid spike; spathe present, but usually very early caducous. Flowers with white, yellow, mauve or bluish-violet petal-like appendages (tepals), rarely with tepals lacking. Stamens with yellowish or brown anthers often early turning blackish; connnective sometimes protruding into a conical appendage; pollen-grains monosulcate, borne in monads. Ovaries green, blue or lilac. Follicles bottle-shaped with a distinct beak, containing 1–12 rounded to fusiform seeds, which are often prominently longitudinally ribbed.

A relatively small genus of about 40 species. All African species are restricted to that continent.

1. Spikes simple; seeds with double testa 2
   Spikes bifid; seeds with simple or double testa 4
2. Leaves with linear blades, not floating on the surface;
   flowers white, unilaterally arranged . . . . . . *1. A. vallisnerioides*
   Leaves with oval to lanceolate blades usually set on a
   narrow petiole and floating on the surface; flowers
   white, pink or mauve, spirally or unilaterally arranged 3

1

3. Tepals 3–6 mm. long, white, caducous; flowers somewhat
   unilaterally arranged; follicles 2–3 mm. long . . .   *2. A. stuhlmannii*
   Tepals 2–4 mm. long, pink or mauve, persistent; flowers
   spirally arranged; follicles 3.5–6 mm. long . . . .   *3. A. afroviolaceus*
4. Plants dioecious; tepals absent on ♀ plants; ♂ plants with
   white tepals . . . .   *4. A. nudiflorus*
   Plants monoecious; tepals always present; separate ♂
   plants absent, or if ovaries reduced then tepals pink
   or lilac 5
5. Tepals yellowish; base of leaf-blade always cordate   *5. A. subconjugatus*
   Tepals pink to lilac, or white; base of leaf-blade seldom
   cordate (if cordate then flowers mauve) 6
6. Leaves usually with a distinct broad floating leaf-blade;
   flowers pinkish or mauve, rarely white (if white then
   leaf-blades 1–3 cm. wide); testa double . . . . .   *6. A. abyssinicus*
   Leaf-blades relatively narrow (in East African plants 3–8
   mm. wide); flowers white; testa simple . . . . .   *7. A. rehmannii*

1. **A. vallisnerioides** *Baker* in Trans. Linn. Soc., Bot. 29: 158 (1875); A. Bennett in F.T.A.
8: 218 (1901); K. Krause in E.P. IV. 13: 19, fig. 9 J–M (1906); F.P.S. 3: 234 (1956); Troupin in
Fl. Garamba 1: 12, fig. 1 (1956); F.W.T.A., ed. 2, 3: 15 (1968); H. Bruggen in B.J.B.B. 43: 197,
fig. 4/1 (1973) & F.A.C.: 3, t. 1 (1974); Lye in U.K.W.F.: 652 (1974); Symoens in Fl.
Cameroun 26: 48, t. 12 (1984); H. Bruggen in Bibl. Bot. 33, 137: 39 (1985). Type: Uganda,
Acholi/Lango Districts, Ukidi, Nov. 1862, *Grant* (K, holo.!)

Plants monoecious. Tuber globular to conical, usually only 5–10 mm. (rarely to 20 mm.)
in diameter; lower half greyish to greenish without roots; upper half with very numerous
roots. Leaves very numerous (often 25–50 or more on robust plants), sessile, ligulate with
obtuse tips, often 10–20 cm. long and 4–8 mm. wide on larger plants, only 5–10 cm. long
and 1.5–4 mm. wide on smaller specimens, without a distinct midrib, but with 7 or 9
indistinct parallel lateral nerves. Flowers white or with a slight violet tinge, drying a warm
yellowish brown, densely set in unbranched unilateral spikes on 5–30 cm. long peduncles
(depending on the depth of the water); spathe 12–18 mm. long, usually persistent when
flowering, caducous when fruiting. Spike usually 2–5 cm. long and ± 10 mm. wide with
tepals spreading, the upper part recurved or spirally twisted; flowers inserted on only one
side of the axis. Tepals 2, white and glossy, rarely with a pinkish or mauve tinge,
particularly below, oval, 3–6 mm. long and usually 2–3 mm. wide, caducous, with several
nerves. Stamens 6, 1.5–2 mm. long; filaments whitish, widened towards the base; anthers
0.4–0.8 mm. long and 0.3–0.5 mm. wide, greyish violet to blackish, but pollen yellow.
Ovaries 3, green, with 6–8 ovules. Follicles green, 4.5–6 mm. long including a ± 1 mm. long
beak. Seeds 1.8–2.5 mm. long and 0.4–0.7 mm. wide, narrowly oblong-lanceolate with
distinct longitudinal ridges (made of the loose outer testa), greenish with the brownish
inner testa visible beneath. Fig. 1.

UGANDA. Karamoja District: Komolo, July 1930, *Liebenberg* 210!; Teso District: Bukedea, 9 May 1970,
Lye & Katende 5360!; Mengo District: Entebbe, shore below Botanic Garden, 17 Jan. 1958, *Lind*
2315!
KENYA. Uasin Gishu District: Kipkarren, 1931, *Brodhurst-Hill* 32!
DISTR. U1, 3, 4; K3; scattered throughout tropical Africa from Senegal and Sierra Leone to Ethiopia
and with a southern limit in Zambia
HAB. In shallow seasonal rock-pools and lake-edges, locally abundant, usually submerged but
sometimes on wet soil; 1050–1700 m.

2. **A. stuhlmannii** *Engl.* in N.B.G.B. 1: 26 (1895) & P.O.A. C: 94 (1895); A. Bennett in
F.T.A. 8: 218 (1901); K. Krause in E.P. IV. 13: 12, fig. 9 N–Q (1906); Peter in Abh. K. Ges.
Wiss. Göttingen 13(2): 40 (1928) & F.D.O.-A. 1: 117 (1929); Obermeyer in F.S.A. 1: 90, fig.
27/2 (1966); Podlech in Prodr. Fl. SW.-Afr. 143: 3 (1966); H. Bruggen in B.J.B.B. 43: 201,
fig. 4/2 (1973) & F.A.C.: 4 (1974); Lye in U.K.W.F.: 652 (1974); H. Bruggen in Bibl. Bot. 33,
137: 40 (1985). Type: Tanzania, Mwanza District, Uzinza [Usinja], Bugando, *Stuhlmann*
3541 (B, holo.!)

Plants monoecious. Tuber globular to conical, 6–12 mm. long and 4–10 mm. in
diameter, with very numerous whitish slender roots. Leaves usually 5–10 per tuber;
petiole 5–20 cm. long and 0.2–0.6 mm. thick; blade narrowly oblong to lanceolate, 2–5 cm.

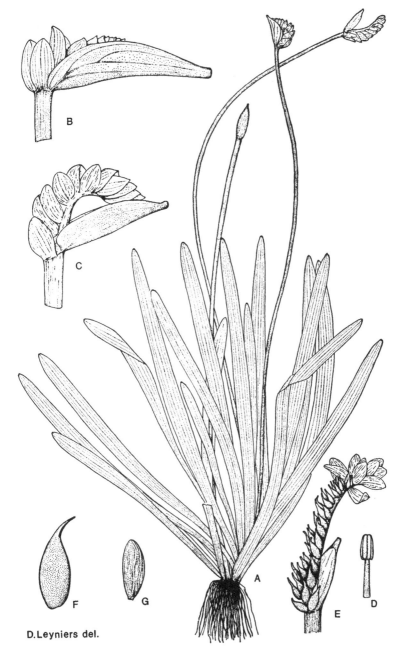

D.Leyniers del.

FIG. 1. *APONOGETON VALLISNERIOIDES* – **A**, habit, × 1; **B, C**, inflorescences, × 3; **D**, stamen, × 10; **E**, infructescence, × 3; **F**, fruit, × 10; **G**, seed, × 10. A–C, from *Troupin* 1585; D, from *César & Ménaut* 167; E–G, from *Boutique* 112. Reproduced from Flore d'Afrique Centrale.

long and 4–10 mm. wide, with 5–9 parallel lateral nerves, floating on the surface of the water; base cuneate to truncate, tip obtuse; primary leaves occasionally produced, 1–2 cm. long, linear, without blade. Flowers white, often drying a bluish white, rather laxly set on unbranched somewhat unilateral spikes on slender peduncles; peduncles 5–15 cm. long and 1–1.5 mm. thick above, green but inflorescence-axis and base white; spathe 6–10 mm. long, greenish, usually early caducous. Spike 1–2 cm. long and 5–10 mm. wide with tepals spreading, the flowers usually inserted on one side of the axis, but at maturity often more cylindrical, only 7–15-flowered. Tepals 2, white and shiny, but often with a lilac tinge, ligulate, 3.5–5.5 mm. long and 1–2 mm. wide, 1-nerved. Stamens 6; filaments 2–2.5 mm. long, lilac but whitish near the thicker base; anthers 0.5–0.6 mm. long and 0.3–0.4 mm. wide, brownish yellow, but turning black when withering. Ovaries 2–4, green with 2–10 ovules. Follicles green to somewhat bluish (but beak whitish at least near the tip), 2–3 mm. long, including a ± 1 mm. long curved beak, and 0.5–0.6 mm. wide. Seeds elliptic, 1.5–2.5 mm. long and ± 0.5 mm. wide, with a double testa, not longitudinally ridged.

KENYA.   Northern Frontier Province: Moyale, 18 Apr. 1952, *Gillett* 12846!; Nairobi National Park, 22 Nov. 1977, *M.G. Gilbert* 4930!; Teita District: Tsavo National Park East, NE. of Manga Hill, Mudanda Rock, 30 Dec. 1971, *Faden, Faden & Smeenk* 71/992!
TANZANIA.   Musoma District: 97 km. from Seronera to Kleins Camp, 6 Apr. 1961, *Greenway & Myles Turner* 9987!; Moshi District: Arusha–Moshi, Sanya plain, 7 Apr. 1966, *Leippert* 6456!; Dodoma District: Kazikazi, 9 Apr. 1933, *B.D. Burtt* 4650!
DISTR.   K1, 4, 7; T1, 2, 5; Zimbabwe, Botswana, Namibia and South Africa
HAB.   Temporary shallow rain-pools, often in rock-pavements; 550–1650 m.

SYN.   *A. gracilis* A. Bennett in Fl. Cap. 7: 43 (1897). Type: South Africa, Transvaal, Houtbosch, *Rehmann* 5761 (Z, holo.!, K, iso.!)

NOTE.   Plants from Zambia described as *A. gramineus* Lye in Norw. Journ. Bot. 18: 190, fig. 2 (1971) are very similar to *A. stuhlmannii* in floral structure, but the flowers are spirally arranged on the axis and the flowers dry a warmer yellowish brown. *A. gramineus* never produces leaves with a widened blade, and is probably adapted to wet poor soils, rather than an aquatic habitat. Whether *A. gramineus* is a species of its own or an infraspecific variant of *A. stuhlmannii*, will have to be decided after more field-studies.

3. **A. afroviolaceus** Lye in Bot. Notis. 129: 68 (1976); H. Bruggen in Bibl. Bot. 33, 137: 50 (1985). Type: Kenya, Kiambu District, Thika, *Lye, Katende & Faden* 6348 (MHU, holo.!, K, iso.!)

Plants monoecious. Tuber oval to elongate or cylindrical, 10–25 mm. long and ± 15 mm. thick, the upper part with numerous whitish roots. Leaves few–many (often 3–15 per plant); petiole usually 5–20 cm. long; blades 2–8 cm. long and 3–16 mm. wide, oval to linear-oblong, with 5–7 parallel main nerves, cuneate to truncate base and blunt or subacute tip, more rarely the blades have a long attenuate base and are then not well separated from the petiole. Flowers mauve to lilac, densely set in unbranched spikes and facing in all directions; peduncle greenish, but often white to mauve below, dull maroon to brown above, 5–30 cm. long (depending on the depth of the water) and 0.5–2.5 mm. thick; mature spathe not seen, very early caducous. Spike 6–30 mm. long and 5–10 mm. wide, 10–30-flowered, sometimes elongating to 40 mm. long when fruiting. Tepals 2, purple or mauve, obovate or spathulate, 2–6 mm. long and 1–1.4 mm. wide, 1-nerved, long-persistent. Stamens 6 or 1 or lacking in apomictic specimens; filaments 1.5–3 mm. long, purple, thickened towards the base; anthers yellow to dark brown, 0.3–0.7 mm. long, the connective sometimes with a flattened conical protuberance ± 1 mm. long. Ovaries 3–8, mauve, with 2–9 ovules. Follicles greenish, 3.5–6 mm. long and ± 1.5 mm. wide including a ± 1.5 mm. long narrow beak which is often dark purple at the tip. Seeds narrowly oblong-lanceolate to cylindrical, often slightly curved, 1–2 mm. long and ± 0.5 mm. wide, with a double testa, longitudinally ridged or winged, at maturity almost blackish.

KEY TO INFRASPECIFIC VARIANTS

Blades 8–16 mm. wide, oval to oblong; tepals 2–4 mm. long;
    follicles 3.5–4.5 mm. long . . . .   var. **afroviolaceus**
Blades 3–6 mm. wide, linear-oblong; largest tepals 4–5 mm.
    long; mature follicles 4–6 mm. long . . . .   var. **angustifolius**

var. **afroviolaceus**

Plant with 5–15 leaves per plant; blades oval to oblong, 8–16 mm. wide, sometimes with a long attenuate base and then not well separated from the petiole. Spike 10–30-flowered. Tepals 2–4 mm. long. Anthers yellow to dark brown, 0.5–0.7 mm. long. Follicles 3.5–4.5 mm. long.

KENYA. Kiambu District: Thika, 1 July 1971, *Lye, Katende & Faden* 6348!
TANZANIA. Mbeya District: Usangu plain near Utengule [Utencile], 28 Jan. 1963, *Richards* 17567! & Usangu plain, 16 km. N. of Rujewa, *Anderson* 1184!; Iringa District: 11 km. SE. of Iringa on Dabaga road, 8 Feb. 1962, *Polhill & Paulo* 1387!
DISTR. **K4; T7**; also in Zambia and Zimbabwe
HAB. In mud in small seasonal streams, rivulets, ditches and drainage channels, or in pools and marshes; 1000–1600 m.
SYN. *A. violaceus* Lye in Norw. Journ. Bot. 18: 187, fig. 1 (1971); H. Bruggen in B.J.B.B. 43: 205, fig. 4/4 (1973) & F.A.C.: 2 (1974); Lye in U.K.W.F.: 650 (1974), *non A. violaceus* Lagerh. (1920). Type: as for *A. afroviolaceus*

var. **angustifolius** *Lye* in Lidia 2: 6 (1988). Type: Kenya, 32 km. from Nairobi on Thika road, *Agnew & Hanid* 7522 (NAI, holo.!)

Plants usually with few (3–5) leaves only; blades 3–6 mm. wide, always well separated from the petiole. Spike 10–20-flowered. Largest tepals 4–6 mm. long, but upper ones sometimes shorter. Anthers dark brown to black, 0.3–0.5 mm. long. Mature follicles 4–6 mm. long.

KENYA. Fort Hall District: Thika hillside N. of Thika R., 7 May 1967, *Faden* 67/312!; Kiambu District: 32 km. Nairobi–Thika, 21 Nov. 1965, *Agnew & Hanid* 7522!
DISTR. **K4**; not known elsewhere
HAB. In mud in small temporary pools; 1400–1600 m.

4. **A. nudiflorus** *Peter* in Abh. K. Ges. Wiss. Göttingen 13, 2: 40, t. 13 c–e (1928), in claves & F.D.O.-A.: 116, t. 7 c–e (1929); Bonstedt in Parey's Blumengärtnerei 1: 107, fig. (1931); H. Bruggen in B.J.B.B. 43: 210, fig. 4/7 (1973); Lye in U.K.W.F.: 650 (1974); H. Bruggen in Bibl. Bot. 33, 137: 56 (1985). Lectotype — see Bruggen (1973): Tanzania, Moshi District, Moshi–Sanya, km. 347.5, *Peter* 41726 (B, lecto.!)

Plants dioecious. Tuber globular, oval or irregularly shaped, brown or yellowish brown, 8–30 mm. long and 8–20 mm. thick, with numerous brownish roots in upper part. Leaves usually many; petiole 5–40 cm. long and 0.5–2 mm. thick, but sometimes wider at the pale or pinkish sheathing base; blade floating on the surface of the water, oblong to oblong-lanceolate (rarely linear-lanceolate), 2–13 cm. long and 0.3–4.5 cm. wide, base cuneate to truncate, tip blunt or acute; midrib distinct and with 2–4 parallel main nerves on each side of the midrib. Flowers white or yellowish white, set in bifid spikes with flowers facing in all directions; peduncles 15–50 cm. long and 0.5–3 mm. thick, greenish, but paler below, not thickened towards the inflorescence; spathes 3–20 mm. long, green to greyish, rather long-persistent. Male spikes relatively laxly flowered, usually 1–8 cm. long and 2–5 mm. wide, consisting of tepals and stamens. Female spikes densely flowered, 1–5 cm. long and 4–8 mm. wide, consisting of ovaries only. Tepals 2 in ♂ flowers, absent in ♀ flowers, white or cream, 1.5–3 mm. long, ligulate, broadly ovate or obovate, 1-nerved. Stamens usually 6, rarely 7–8, ± as long as the perianth; filaments white, hardly widened towards the base; anthers green to yellow, turning blackish with age. Ovaries 3, greenish, with 3–6 ovules. Follicles green to brown, 3–4 mm. long and 1–2 mm. wide including a 1–1.5 mm. long narrow beak. Seeds 1.5–2 mm. long and 0.4–0.7 mm. wide, greyish with paler longitudinal ribs; testa double.

KENYA. N. Frontier Province: 2 km. N. of El Wak, 30 Apr. 1978, *M.G. Gilbert & Thulin* 1268!; Baringo District: Lake Kamarr [Kamar], 26 Oct. 1961, *Edmondson* 35!
TANZANIA. Masai District: Great North Road, Tarangire, 8 May 1962, *Polhill & Paulo* 2388!; Mbulu District: 10 km. from Lake Manyara, near Kwa Kuchinja [Kuchinia], 24 Nov. 1965, *Leippert* 6160! & Msitu Watembo, 13 Apr. 1968, *Ludanga* in *Mweka* 2390!
DISTR. **K1, 3; T2, 3**; eastern Ethiopia and Somalia
HAB. In shallow lakes, seasonal pools, ponds and rivers; 420–1200 m.
SYN. *A. nudiflorus* Peter var. *angustifolius* Peter in Abh. K. Ges. Wiss. Göttingen 13(2): 40, t. 13/f, g (1928), in claves & F.D.O.-A. 1: 116, t. 7/f, g (1929). Type: Tanzania, Lushoto District, Pangani R., NW. of Buiko, *Peter* 40967 (B, holo.!)
   [*A. rehmannii* sensu Peter in F.D.O.-A: 116–117 (1929), *non* Oliv.]
   [*A. natalensis* sensu Peter in F.D.O.-A: 116–117 (1929), *non* Oliv.]
   [*A. leptostachys* sensu Peter in F.D.O.-A: 116–117 (1929), pro parte (only for *Peter* 11102 & 11387), *non* Engl. & K. Krause]

5. A. subconjugatus *Schumach. & Thonn.*, Beskr. Guin. Pl.: 183 (1827); A. Bennett in F.T.A. 8: 217 (1901); F.P.S. 3: 233, fig. 63 (1956); Berhaut in Fl. Sénégal, ed. 2: 304, 315, fig. p. 317 (1967); F.W.T.A., ed. 2, 3: 15, fig. 321 (1968); H. Bruggen in B.J.B.B. 43: 211, fig. 4/8 (1973); Lye in U.K.W.F.: 650 (1974); Symoens in Fl. Cameroun 26: 51, t. 13 (1984); H. Bruggen in Bibl. Bot. 33, 137: 58 (1985). Type: Ghana, *Thonning* 109 (C, holo.!)

Plants monoecious. Tuber oval or irregularly shaped, 10–40 mm. long and 10–30 mm. thick, densely set with brown and whitish roots above. Leaves few–many; petiole 5–60 cm. long and 1–6 mm. thick, but often wider at the sheathing base; blade oblong to oblong-lanceolate, 5–23 cm. long and 1.5–6 cm. wide, base prominently cordate, apex obtuse or rounded; main parallel lateral nerves 3–5 on each side of the midrib; cross-nerves prominent on lower surface. Flowers cream to yellow, densely set in bifid spikes with the flowers facing in all directions; peduncles 10–70 cm. long and 2–7 mm. thick, often slightly thicker below the inflorescence; spathes 20–50 mm. long, including an up to 20 mm. long acumen, green, early caducous. Spikes bisexual, dense-flowered, 15–50 mm. long and 4–8 mm. wide when flowering, elongating to 130 mm. long and 13 mm. wide when fruiting. Tepals usually 2, cream to yellow with a slightly darker midrib and brownish gland-dots, ovate to ligulate, 1.5–3 mm. long, often persistent. Stamens usually 6; filaments 2–3 mm. long, yellow, slightly widened towards the base; anthers 0.3–0.5 mm. long and 0.3–0.4 mm. wide, yellow, not turning black with age. Ovaries 4–6, green; ovules 10–12. Follicles green, 4–6 mm. long and 2–3 mm. wide including a 1–2 mm. long beak. Seeds 1.4–2 mm. long and 0.5–0.9 mm. wide, cylindric, greyish brown with longitudinal pale wings; testa double.

UGANDA. Karamoja District: W. of Toror, Kotido, 3 June 1940, *A.S. Thomas* 3696!
DISTR. U1; only known from one collection from East Africa, more widespread in West Africa from Senegal, Mali, Ghana, Nigeria and Cameroon to Chad, possibly even in the Sudan
HAB. In temporary or possibly permanent pools and swamps; 1200 m.

SYN. *Ouvirandra heudelotii* Kunth in Enum. Pl. 3: 593 (1841). Type: Senegal, Oualo, *Heudelot* 433 (P, holo.!, K, iso.!)
Aponogeton heudelotii (Kunth) Engl. in E.J. 8: 271 (1886); K. Krause in E.P. IV. 13: 15, fig. 3/A–D (1906)

6. A. abyssinicus *A. Rich.*, Tent. Fl. Abyss. 2: 351 (1851); A. Bennett in F.T.A. 8: 218 (1901); E.P.A.: 1202 (1968); H. Bruggen in B.J.B.B. 43: 221, fig. 7/12 (1973) & F.A.C.: 9, photo. facing p. 1 (1974); Lye in U.K.W.F.: 650, fig. p. 651 (1974); H. Bruggen in Bibl. Bot. 33, 137: 60 (1985); Lye in Lidia 1: 67 (1986). Type: Ethiopia, Tigray, near Aksum [Axum], *Schimper* 1483 (P, holo.!, BR, GOET, K, M, iso.!)

Plants monoecious. Tuber oval, shortly cylindrical or irregularly shaped, 10–30 mm. long and 5–20 mm. thick, densely set with whitish roots above. Leaves few–many; petiole (except in var. *graminifolius*) 3–60 cm. long and 0.6–3 mm. thick, greenish, terete or somewhat compressed; blade usually oval to lanceolate, floating, with acute to obtuse tip and cuneate to truncate base, very rarely (only in var. *cordatus*) the base is prominently cordate, 3–12 cm. long and 0.5–5 cm. wide, green, but sometimes dark-spotted beneath; midrib very prominent on lower surface, parallel nerves 2–4 on each side of the midrib; very rarely the leaves sessile, linear and grass-like (in var. *graminifolius*). Inflorescence a bifid spike 10–50 mm. long and 2–8 mm. wide when flowering, with many usually densely set bisexual or ♀ flowers facing in all directions; peduncles 5–45 cm. long and 1–5 mm. thick, terete or somewhat flattened, green to reddish; spathe 5–20 mm. long, pale green with darker longitudinal lines (and sometimes spots). Flowers usually consisting of 2 tepals, 6 stamens and 3 ovaries (rarely no stamens and up to 7 ovaries in apomictic plants). Tepals persistent or caducous, pinkish, mauve to bluish violet, rarely white (only in var. *albiflorus*), oval, obovate to ligulate, or with a narrow base and a wider upper part, 1–5 mm. long, with 1 nerve, rarely densely set with numerous reddish brown gland-dots (only in var. *glanduliferus*). Stamens 2–3 mm. long; filaments white or violet, somewhat thickened towards the base; anthers 0.2–0.4 mm. long and 0.2–0.3 mm. wide, violet to almost black, but pollen yellow. Follicles grey to brownish, 3–7 mm. long with a prominent 1–2 mm. long beak of the same colour as the remainder of the follicle or much darker, containing 4–14 ovules. Seeds elliptic to elongate, 0.7–1.5 mm. long and 0.2–0.4 mm. wide, with longitudinal ridges, but not prominently winged; testa double.

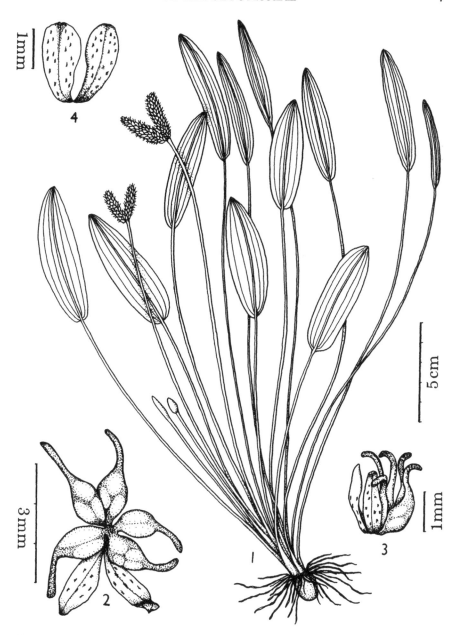

FIG. 2. *APONOGETON ABYSSINICUS* var. *ABYSSINICUS* — **1**, habit; **2, 3**, flower; **4**, tepals. All from *Verdcourt & Fraser Darling* 2294. Drawn by Gerd Mari Lye.

KEY TO INFRASPECIFIC VARIANTS

1. Leaves sessile, linear, without a floating blade; only in
   southern Tanzania . . . .   e. var. **graminifolius**
   Leaves with a distinct petiole and an ovate to lanceolate
   floating blade 2
2. Flowers white; tepals 1–2 mm. long; sea-level to 600 m.   b. var. **albiflorus**
   Flowers pinkish, mauve or violet, tepals 1–5 mm. long;
   mostly at higher altitudes 3
3. Leaf-base prominently cordate; tepals 1–2 mm. long   c. var. **cordatus**
   Leaf-base truncate to cuneate; tepals mostly 2–5 mm. long 4
4. Mature tepals 2–2.5 mm. long, prominently gland-dotted . d. var. **glanduliferus** .
   Mature tepals 2–5 mm. long, not gland-dotted . . . .   a. var. **abyssinicus**

a. var. **abyssinicus**; Lye in Lidia 1: 71, figs. 1, 5 (1986)

Slender to robust aquatic with a small to large tuber. Leaf-blades oval to lanceolate, 3–12 cm. long and 0.5–5 cm. wide, with truncate or cuneate base; petiole 3–30 cm. long. Spike 1–5 cm. long and 3–8 mm. wide; peduncles slender to fairly robust (1–4 mm. thick). Tepals oval to ligulate, 2–5 mm. long, pinkish to bluish violet, membranous, without gland-dots. Follicles 3–7 mm. long, green or greyish, with a prominent 1–2 mm. long beak not conspicuously darker. Seeds 4–12, elongate, 0.7–1.5 mm. long and 0.2–0.4 mm. wide. Fig. 2.

UGANDA. Karamoja District: Kotido, Apr. 1960, *J. Wilson* 1893!; Toro District, Lake George, Kashenyi, 8 Nov. 1968, *Lock* 68/284!; Ankole District: Mbarara, Rushozi [Roshoshi] dam, 7 Nov. 1954, *Lind* 506!

KENYA. Naivasha District; 10 km. NW. of Naivasha, 8 Apr. 1969, *Lye* 2454!; Machakos District: near Athi R., 14 May 1961, *Polhill* 421!; Masai District: Mara Plains, Keekorok [Egalok], 21 Oct. 1958, *Verdcourt & Fraser Darling* 2294!

TANZANIA. Musoma District: Klein's Camp to Seronera, ± 48 km. from Seronera, 6 Apr. 1961, *Greenway & Myles Turner* 10005!; Lushoto District: Umba steppe, Kivingo, 28 Dec. 1929, *Greenway* 1978!; Mpanda District: Katisunga, 6 Jan. 1950, *L. Thomas* 74!

DISTR. U 1, 2; K 1, 3, 4, 6; T 1, 3,5; scattered through eastern Africa from Ethiopia to southern Tanzania and Shaba in Zaire; a very poor plant from Malawi possibly also belongs here

HAB. In temporary pools, waterfilled ditches, dams and other seasonal gatherings of water, usually in 5–40 cm. deep water; 440–1800 m., but rarely found below 800 m.

SYN. *A. leptostachyus* Engl. & K. Krause var. *minor* Baker in Trans. Linn. Soc., Bot. 29: 158 (1875). Type: Tanzania, Dodoma District, Uyanzi, Mgongo Thembo, 24 Jan. 1863, *Grant* (K, holo.!)
Ouvirandra hildebrandtii Eichler in Sitz. Ges. Nat. Freunde Berlin: 193 (1878) & Monatschr. Ver. Bef. Gartenb. Kgl. Pr. St. 22: 6, t. 1 (1879). Type: Kenya, Kitui District, Ukamba, *Hildebrandt* 2645 (B, holo. †, K, M, iso.!)
Aponogeton boehmii Engl. in N.B.G.B. 1: 26 (1895); A. Bennett in F.T.A. 8: 218 (1901); Engl. & K. Krause in E.P. IV. 13: 14, fig. 3K–N (1906); Peter in Abh. K. Ges. Wiss. Göttingen 13(2): 40 (1928) & in F.D.O.-A. 1: 116–117 (1929). Type: Tanzania, Tabora District, Unyamwezi, Wala R., *Böhm* 98 (B, holo.!, Z, iso.!
A. leptostachyus Engl. & K. Krause var. *abyssinicus* (A. Rich.) Engl. & K. Krause in E.P. IV. 13: 14 (1906)
A. braunii K. Krause in E.J. 57: 240 (1921). Type: Tanzania, Bukoba District, Msera, *Braun* in Herb. Amani 5494 (EA, holo.!, B, iso.!)
A. oblongus Peter in Abh. K. Ges. Wiss. Göttingen 13(2): 40, t. 13/a, b (1928) & in F.D.O.-A. 1: 116–117 (1929) & F.D.O.-A., Anhang: 9 (1938). Types: Tanzania, Tabora District, Ngulu, Malongwe railway bridge, *Peter* 34581 & 39208 (both B, syn. †)
A. leptostachyus sensu Peter in Abh. K. Ges. Wiss. Göttingen 13(2): 40 (1928), pro parte, & in F.D.O.-A.: 116–117 (1929), pro parte, *non* Engl. & K. Krause]

b. var. **albiflorus** Lye in Lidia 1: 73, fig. 2 (1986). Type: Tanzania, 5 km. S. of Bagamoyo on road to Dar es Salaam, Bako Swamp, *Wingfield* 2185 (K, holo.!, DSM, EA, iso.!)

Slender to fairly robust aquatic with an oval tuber. Leaf-blades oval to lanceolate, 3–10 cm. long and 1–3 cm. wide, with a truncate to slightly cordate base; petiole 5–30 cm. long. Spike 2–4 cm. long and 3–7 (–9 in fruit) mm. wide; peduncles 1–2 mm. thick, greenish. Tepals oval-elliptic to ligulate, 1–2 mm. long, membranous, white, without gland-dots. Follicles 3–4 mm. long, grey to brownish, the beak not usually conspicuously darker. Seeds many, ± 0.8 mm. long and 0.2–0.3 mm. wide, elongate.

KENYA. Teita District: Tsavo National Park East, NE. of Manga Hill, Mudanda Rock, 7 May 1974, *R.B. & A.J. Faden & Kingston* 74/544!
TANZANIA. Bagamoyo District: Bako Swamp, 3 June 1973, *Wingfield* 2185!, Kilwa District: Selous Game Reserve, Kingupira, 28 Mar. 1976, *Vollesen* 3387!
DISTR. K 7; T 6, 8; not known elsewhere
HAB. Rice-fields and temporary pools and waterholes; near sea-level to 600 m.

c. var. **cordatus** *Lye* in Lidia 1: 75, fig. 3 (1986). Type: Kenya, Northern Frontier Province, Mathews Range, Olkanto, *J. Adamson* in *Bally* 4354 (K, holo.!, EA, iso.!)

Slender to robust aquatic with a globular to oval edible tuber. Leaf-blades oval, 3–10 cm. long and 1.5–5 cm. wide, with strongly cordate base and usually 3 weak parallel lateral nerves on each side of the midrib; petiole most often 10–25 cm. long. Spike 1–5 cm. long and 2–5 (–8 in fruit) mm. wide; peduncles slender to fairly robust, 1–3 mm. thick. Tepals 1–2 mm. long, with a narrower base and wider tip, membranous, mauve to purplish, without gland-dots. Follicles 3–4 mm. long, grey to olive-brown, the beak not conspicuously darker. Seeds up to 14, elongate, 0.8–1 mm. long and 0.2–0.3 mm. wide.

KENYA. Northern Frontier Province: Mathews Range, Olkanto, 15 Dec. 1944, *J. Adamson* in *Bally* 4354!; Tana River District: Kurawa, 7 Oct. 1961, *Polhill & Paulo* 622!
DISTR.   **K** 1, 7; Somalia
HAB.   Temporary pools; 15–900 m.; only in regions with a very low rainfall

d. var. **glanduliferus** *Lye* in Lidia 1: 77, fig. 4A–D (1986). Type: Tanzania, Tabora District, Wala R., S. of Tabora, *Lindeman* 783 (K, holo.!)

Robust aquatic herb with floating blades on 30–50 cm. (or more) long petioles (depending on depth of water); blades elongate, 9–13 cm. long and 2–2.5 cm. wide, with cuneate bases and 3 prominent parallel lateral nerves on each side of the midrib. Spike 3–7 cm. long and 4–6 mm. wide; peduncles 3–5 mm. thick. Tepals obovate, 2–2.5 mm. long, relatively thick, mauve when fresh, but drying greyish brown and densely set with prominent reddish brown gland-dots. Follicles 3–4 mm. long, olive to brown with a very dark (blackish) beak. Seeds ± 6, elliptic, 0.6–0.8 mm. long and 0.3–0.4 mm. wide.

TANZANIA. Tabora District: Wala R., S. of Tabora, 26 May 1940, *Lindeman* 783!
DISTR.   **T** 4; only known from the type collection
HAB.   Pools in river; 1200 m.

NOTE.   It is possible that var. *glanduliferus* is related to *A. subconjugatus*. These two taxa share the characters of glandular tepals as well as a micropapillate surface of the seeds (only apparent with a scanning electron microscope). However, the only fruits of var. *glanduliferus* known are very immature, so too much weight cannot be put on this character alone. New and mature material of this taxon is consequently highly desired. Possibly this plant may prove to be a distinct species.

e. var. **graminifolius** *Lye* in Lidia 1: 79, fig. 4E, F (1986). Type: Tanzania, Ufipa District, Kasisiwue plain, *Richards* 10352 (K, holo.!)

Slender grass-like herb with an oval-cylindrical tuber 5–8 mm. in diameter. Leaves sessile, linear, 2–6 cm. long and 1–2 mm. wide, without a floating blade. Spike 1–1.5 cm. long and 4–6 mm. wide; peduncles 1–2 mm. thick, reddish or purplish. Tepals elliptic-lanceolate to ligulate, 2–3 mm. long, bluish purple, without gland-dots. Fruit not seen.

TANZANIA. Ufipa District: Kasisiwue plain, 15 Dec. 1958, *Richards* 10352!
DISTR.   **T** 4; only known from the type collection
HAB.   Shallow water in boggy part of grassland; 1650 m.

7. **A. rehmannii** *Oliv.* in Hook., Ic. Pl. 15, t. 1471b (1884); A. Bennett in Fl. Cap. 7: 44 (1897) & F.T.A. 8: 217 (1901); Engl. & K. Krause in E.P. IV. 13: 15 (1906); Podlech in Prodr. Fl. SW.-Afr. 143: 2 (1966); H. Bruggen in Bibl. Bot. 33, 137: 63 (1985). Type: South Africa, Transvaal, between Kleinsmit and Kameelpoort, *Rehmann* 4835 (K, holo.!, Z, iso.!)

Plants monoecious. Tuber oval or irregularly shaped, 15–40 mm. long and 10–25 mm. thick, densely set with white or brownish roots in upper part. Leaves few–many (usually 5–10 fresh leaves per plant); petiole 3–30 cm. long and 1–3 mm. thick, usually prominent, but occasionally soft juvenile linear leaves without stalk and blade are produced; blade usually floating on the surface of the water, oval to linear-lanceolate, 3–10 cm. long and 0.3–2.5 cm. wide (East African plants have 3–8 mm. wide blades), with apiculate or subacute tip and cuneate or somewhat rounded base, sometimes strongly attenuate; midrib prominent and with 2–3 weak parallel nerves on each side of the midrib. Flowers white, densely set in bifid spikes with the flowers facing in all directions; penduncles 5–30 cm. long and 0.5–2 mm. thick, green to olive-brown with white or purplish base, not thickened below the inflorescence; spathe up to 10 mm. long, very early caducous. Spikes bisexual, densely-flowered, 0.5–2.5 cm. long and 4–6(–9 in fruit) mm. wide. Tepals 1–2, white or with a faint pinkish tinge, ovate or obovate, 2–4 mm. long and 0.7–2 mm. wide, 1-nerved. Stamens 6, but often lacking since the plants are often apomictic. Ovaries 3–6, grey to purplish when young; ovules 2–6. Follicles 5–6 mm. long and ± 3 mm. wide, greyish brown with whitish beak. Seeds cylindrical, straight or slightly curved, 2.5–3.8 mm. long and 0.8–1.2 mm. wide, greyish green, without longitudinal ridges; testa simple.

KENYA. Fort Hall District: N. side of Chania [Chanya] R., Thika, 1 July 1971, *Lye, Katende & Faden* 6375! & hillside N. of Thika R., 7 May 1967, *Faden* 67/313!
DISTR. **K** 4; Zambia, Zimbabwe, Mozambique, Botswana, Namibia and South Africa, but not yet found in Tanzania
HAB. Seasonal pools in rock pavement, or other temporary gatherings of water; 1450 m.

SYN. *A. hereroensis* Schinz in Bull. Herb. Boiss., sér. 2, 1: 764 (1901). Types: Namibia, pool E. of Windhoek, *Dinter* 589 (Z, syn.!) & N. of Waterberg, 10 Apr. 1899, *Dinter* s.n. or 589a (Z, syn.!)
A. rehmannii Oliv. var. *hereroensis* (Schinz) Engl. & K. Krause in E.P. IV. 13: 16 (1906)
A. junceus Schlechtend. subsp. *rehmannii* Oberm. in Fl. Pl. Afr. 37, t. 1449 (1965); H. Bruggen in B.J.B.B. 43: 229, fig. 7/15 (1973) & F.A.C.: 10 (1974)
[*A. junceus* sensu Oberm. in F.S.A. 1: 88, fig. 26a (1966); Lye in U.K.W.F.: 650 (1974), *non* Schlechtend.]

NOTE. *A. rehmannii* is closely related to *A. junceus*, but it differs in producing floating leaf-blades and having the flowers facing in all directions. It is here regarded as sufficiently distinct to warrant specific recognition although included in *A. junceus* in F.S.A. *A. rehmannii* is the only East African species of the genus with simple testa.

INDEX TO APONOGETONACEAE

T - #0147 - 101024 - C0 - 234/156/1 [3] - CB - 9789061913412 - Gloss Lamination